Solomon Heydenfeldt

The Unison of the Conscious Force

Electro-magnetism and hypnotism. An outline of the secret of the

Buddhists.

Solomon Heydenfeldt

The Unison of the Conscious Force
Electro-magnetism and hypnotism. An outline of the secret of the Buddhists.

ISBN/EAN: 9783337240783

Printed in Europe, USA, Canada, Australia, Japan

Cover: Foto ©berggeist007 / pixelio.de

More available books at **www.hansebooks.com**

OF THE

CONSCIOUS FORCE.

To the Medical Profession.

ELECTRO-MAGNETISM

AND

HYPNOTISM.

An Outline of the Secret of the Buddhists.

*Theosophy — Spiritualism — Eavesdropping—Espionage—
The Doctrine of Secrets—Secret or Cystic Punishment
—Power of Magnetizers—Semi-Subjectiveness—
Telepsychological Action—Thought Transference.
Mind Reading—Muscle Reading—Transmission by Sight—
Visualizing—Dreams—Dangers of Magnetism—Victim
System—Hypnotism and Magnetism—Appendix—
Distributing Mediums—Etc . Etc.*

By S. HEYDENFELDT, Jr.

SAN FRANCISCO:
W. M. Hinton & Co., Printers and Publishers,
536 Clay Street,
1890.

THE UNISON

CONSCIOUS FORCE.

(A Force Which Carries the Power of Thought.)

ELECTRO-MAGNETISM.

(USE OF THE TERM.)

The reasons for using the term electro-magnetism. instead of magnetism, are many. In the first place " magnetism " has been used synonymously with " mesmerism;" even writers of some authority have confounded the meaning of the terms. Furthermore the various phenomena produced by magnetizing the human body by electricity have been so frequently ascribed to animal magnetism, or to the effect of one mind upon another, that recently it has scarcely ever been used to mean anything else; so few have had explained to them the effects which can be produced by electricity.

The word " *electro* " at once indicates that it is some form of electricity generated either by friction or chemicals.

I did not use the term electricity for the reason. that a body of spiritualists construed it to mean spiritualism, meaning that all the phenomena unexplained, produced by its application to the living body. were due to spiritism.

Again, electro-magnetism is particularly a proper term to use, as it includes the two important factors which produce the results; and notwithstanding that there is a definite application and meaning of the term in mechanical electricity, there is no reason why it should not have another application, when used in connection with living beings, in the manner that magnetizing has its meaning in mechanical forces as well as in the force of life.

UNISON OF THE CONSCIOUS AND SENSORY FORCES.

The brain of one person can be connected with that of another by a current of electro-magnetism generated from an electro-galvanic or Faradic or other battery, which will continue so long as the current is being generated by the battery, passing through the one applying a battery to himself (or having it applied by an operator) to the other, who need not have applied to him either pole of the battery, the earth's magnetism, the moisture, or serum, and the electricity and magnetism in the living body of the latter being sufficient to continue the current once induced for a very considerable period of time. Other parts of the body may, in a like manner, be connected by such a current of electro-magnetism.

After a current of electro-magnetism generated from a battery has connected one person with another or others, certain substances and ethers, some of which will be mentioned further on, may be conducted with the current applied to one of them in the way that chemicals can be introduced into the system by the galvanic cur-

rent. The substance, passing with the current of electro-magnetism from the battery of the person using it
on himself on applying it to another, and being induced
to the others, has the effect of producing a continuing
invisible current, passing from one to the others, and
continuing as long as the current of electro-magnetism
is generated, and for a considerable time after the use
of the battery has been suspended. The substance introduced into the systems of each, together with the
serum, electricity and magnetism generated by the bodies, and the earth's magnetism, have the effect to cause
the current to continue to pass from one to the other.
How long such a connection will continue, by the natural electricity and magnetism generated in the bodies
connected, will of course depend upon the amount of
the substance which has been used, as well as the distance between the persons.

All of the muscles, tissues, fibres and nerves of one
person can be connected with the corresponding ones
of another. Where such a complete connection is
made it follows that the natural electricity and magnetism of each aids in sustaining the current.

Let us suppose that A and B have been thus connected by such a current of electro-magnetism carrying
with it one of these substances. A represents the person who uses the electro-motive force from the battery,
or on whom it is used by the operator. B represents
the person connected with A to whom the current is
induced.

The quantity of the electro-motive force applied to
the person of A will, if sufficient, and the distance of B

be not too great, carry or conduct with it a portion of the magnetic nerve force, muscular force and sensory forces of A to B to such an extent that the movements of B can be felt by A. Any cauterization of A can be felt by B. Any irritation of the skin of B can be felt by A. If the current induced from A to B passes through corresponding muscles of the bodies of A and B, any irritation of the epidermis of the parts connected of either will be felt by the other.

The sense of touch as well as of the muscular movements of B can be detected by A.

If the current from the battery is made to pass through the cerebrum and cerebellum of A to the brain of B, the distance not being too great, and the quantity of the electro-motive force being sufficient, a portion of what may be termed the conscious force of A will be transferred or induced to B, to such an extent that not only the thoughts and words may be communicated from one to the other. but articulated speech is rendered possible, and intelligent conversation may be had by reason of these forces being induced or carried with the current of electro-magnetism to B. B will know what A wishes to say, and express it for him, and, if it is a question, can answer it. If the quantity of electro-motive force applied to A be increased, and is made to pass through his neck and head, the induced forces to B will be sufficient to compel B to express, by articulated speech, whatever A wishes to say, whether B wishes to or not; in such case the muscles of B must be controlled by A or by the person operating on A, by using and applying the electrodes and forceps to the proper muscles. In the same way the operator can force A to express and say whatever he (the operator) pleases by fixing his mind upon the words mentally formed, which are conducted through his muscles and · nerves to the electrode or forceps attached, and thence to A, who can be made to express them either mentally or by inarticulated or articulated expression; and the current, passing to the corresponding muscles of B, forces B to repeat what A said. (See paragraph on " Muscle Reading.")

To such an extent may part of this conscious force, as well as the portions of the sensory forces of A, be , induced to B, that they exist actively and independently in B. The current properly and carefully directed through and from the optic nerves of A to the optic nerves of B, will enable A to read through the eyes of B.

The olfactory nerves are particularly sensitive under this condition of electro-magnetism. A and B being connected, and the olfactory nerves being partly in unison, the slightest odor can be detected by either and both; thus, ammonia inhaled by A can be easily detected by B; also, certain gases and perfumes.

The auditory nerves of both A and B are also particularly sensitized under this condition of electro-magnetism. This will be discussed further on.

Where the electro-magnetism and other forces are induced from A to B, by the application of the electro-motive force to the whole body of A, so much of the conscious force, the magnetic-nerve force, the muscular force and the sensory forces of A may be induced to B, that the condition of A in B has been termed a condition of being incysted, meaning partially incysted. .

While this unison of the conscious force and other forces of A and B exists, and A is partially incysted in B, A and B are not incapacitated from pursuing their usual avocations, although the muscular movements of B may be measurably interfered with.

Of course, a condition of A being so incysted in B, might exist so as to unfit A from attending to his usual duties, from debility caused by absence of those portions of his vital forces which have been temporarily induced to B.

In order to effect as complete a unison as possible, the two persons should be of uniform size, which being impossible, a smaller person should be incysted in a larger one, so that as many nerve fibres of B may receive the corresponding nerve fibres of A. In this way the function of every organ can be felt to be united.

When A is incysted in this way in B, A being a smaller person, B sometimes feels as though there was a film—I will call it a film of magnetism for the lack of a better term—in various parts of his body; its exact distance from the epidermis varies in different parts of the body. I attribute this greater unison to the use of an electro-magazine applied to the whole of the body of A after the unison has been effected by the current from the battery. In this latter statement I may, however, be mistaken; but a more complete unison can be effected by which such a film can be felt by B.

To effect this unison much depends upon the distance of the persons, the quantity of electro-motive force, a proper direction of the current of electro-mag-

netism to the corresponding parts of the bodies of the
persons.

If a larger person is connected with a smaller one,
and an attempt is made to incyst all of his corresponding muscular and nerve fibres, it will cause a sense of
fullness and suffocation in the smaller one, owing to
the larger one having more nerve fibres which seek to
find a place in the corresponding nerve fibres of the
smaller person.

In order to accurately describe the condition of
being incysted, meaning partially incysted, it is necessary to have a thorough knowledge of physiology and
electro-therapeutics, which I have not.

THE CONSCIOUS FORCE.

A FORCE WHICH CARRIES THE POWER OF THOUGHT.

I have used the term conscious force as being a
force which carries the power of thought, for the reason that the vital forces of A can be so incysted in B,
and A be in such a condition that he seems to exert
this power in B when a sufficient quantity of electro-
motive force is applied to his brain, and part of its
forces are induced to the brain of B. It may be possible that this power can, under conditions of great
unison, be exercised by A using that portion of his conscious force which is induced to B to a greater extent
than he can use that which remains in his own brain.

Those who have studied the actions of hypnotics
and the operations of the minds of the subjectives or
semi-subjectives, when partially incysted in others,
may throw some light upon the question as to

whether the conscious force is exercised more in the person in whom they are incysted, or in their own brains; the observation and the reasoning faculties of persons in such condition have probably been studied by thoughtful men.

For the purpose of effecting this union of the conscious and other forces, the persons sought to be connected, should have the current passed through their brains, respectively, with sufficient force to overcome the *resistance of the power of the will*, which will render them temporarily unconscious. This should be resorted to, if a long-continued, constant current passing through the heads of both has failed. It may be possible that when the will power is overcome by hypnotism or mesmerism, or by an anæsthetic or ether, the union may be readily made.

In making a complete connection it must be remembered that the nerve is always in the condition of a closed circuit. The peri-polar arrangement of the molecules must also be borne in mind; also, the effect of applying the polarizing current of the battery by which the positive zones of the nerves will turn toward the negative, and the negative toward the positive; use the electrodes on the two persons in such a manner that the forces of one can be induced to the other.

Galvanic or Faradic or medical coil batteries with zinc carbon cells are used to effect this unison of the conscious and sensory forces.

In order to continue the connection from brain to brain different substances are used, depending on the distance between the persons.

Cerium (atomic weight, 141.2; Roscoe). The oxa-
late and the oxide; sulphuric acid; gypsum; natron;
magnesia; gum Arabic and other hydro-carbons.

Many other chemicals are said to be used which
must be made the subject of experimentation.

Substances, chemicals, food, etc., which generate
serum, nerve force and magnetism, are administered to
one or more of the persons connected, besides using
acids and ethers by electro-medication.

HEARING AND TRANSMITTING.

Persons under some conditions of electro-magnetism become transmitters of sound. What they hear is
transmitted to others, whose auditory nerves are sensitized and who are in a line of magnetism. *The act of
hearing is in itself a repetition.*

CONFIDENTIAL COMMUNICATIONS.

Transmitting what is heard, being the case, a person
in such condition should not receive confidential communications. Some persons, perhaps, are *employed* to
seek confidential communications and transmit them to
persons who record them, or, being in such condition
and not knowing it, or not knowing the truth of the
statement that " the act of hearing is a repetition in
itself," might be the innocent means of transmitting
what was spoken confidentially.

EAVESDROPPING.

At not too great a distance, persons whose auditory

nerves are sensitized can act as eavesdroppers and stenograph what is transmitted.

When a person in this condition hears what is said, it may be transmitted to one or more eavesdroppers who are in a line of magnetism. This line of magnetism depends greatly upon the person hearing, and the eavesdroppers not moving their position during the conversation. A number of eavesdroppers being in the neighborhood one or more of them by attention may place himself in the line of magnetism and hear and transcribe what is said. The eavesdroppers can meet, compare their manuscripts and in this way succeed in procuring a tolerably correct engrossed manuscript embodying the conversation.

THE LINE OF MAGNETISM.

Persons who are in a condition of electro-magnetism, whose brains are not connected, may at not too great a distance hear what is said to a transmitter. The term "line of magnetism" used in this connection may be illustrated in the same way that persons hear who are not in any condition of electro-magnetism. If a person is in a large room he may hear when in one position and not in another; if the person speaking is at some distance from him the position of the speaker it may require the person spoken to to change his position in order to hear with facility.

Again, I may illustrate it by two persons being out of doors at a considerable distance apart; the person spoken to changes the position of his head in order to hear what is said until he finds the line or current of the vibrations from the mouth of the speaker; if he is straining himself to hear what is said he does not move his position; if he does, the voice but not the words reach him.

Where the brains of one or more persons are *slightly* connected by a current of electro-magnetism the line of magnetism may be illustrated by an invisible wire connecting each with the other, which must extend from the auditory nerves of one to the other or others. By movement this line may be broken, and may be recovered again by change of position and attention.

Persons who are in a condition of electro-magnetism hear the sounds of the voice of human beings who are in a similar condition by the inner ear direct, and also the sounds which they transmit.

The brain of the person who is directly receiving the confidential communication, being connected with the brain of another, or others. transmits to him. or them, with more certainty all that he hears.

The person might be unaware of the fact that what he heard was transmitted, and he might also even be unaware of the fact that his brain was connected with another, or others, by a current of electro-magnetism. or knowing that his brain was connected with another, he might believe that such other was the only person with whom his brain was connected.

He might listen to a private and confidential communication, thinking it would be transmitted to another by agreement of the three, and at the same time unconsciously be transmitting by such brain connection to many others.

A person, in order to insure a most accurate record of what he hears, can repeat what was said to him by a slight, unobserved, inarticulated expression, using the tongue and the muscles of the mouth. like most peopl e do when reading to themselves.

THE DOCTRINE OF SECRETS.

A person can be electro-magnetized without know-ing it. He can be drugged or chloroformed, and sub-jected to a current of electro-magnetism and his brain connected with another or others.

Such a condition can probably be brought about by passing a light, gentle current through him while asleep, without detection, by persons in adjoining rooms with proper electrical appliances. Once in this condi-tion, without knowing it, with the brain connection affected, his every thought and every act and every motive could be known.

He would become unconsciously a transmitter of confidential communications, and aid in a system of espionage by which he might place himself or his friends in the power of others by their secrets becom-ing known in this way.

IN THE POWER OF MAGNETIZERS.

A person .so electro-magnetized, not knowing his condition, is physically in the power of the magnetizers. His muscular movements can be interfered with. Se-vere pains can be produced in all parts of the body.

In fact, whatever effect that can be produced by the application of an electric current, whether from the frictional, Faradic or galvanic battery, directly applied to the whole body of a person, or whether by localized application, corresponding effects can be produced upon another whose muscles, tissues, fibres and nerves are connected with the person to whom the current is applied, provided the distance is not too great and the quantity of electro-motive force is sufficient.

The action of the heart can be suspended and death ensue.

Poisons to produce fatal results, can be introduced into his system through the person partially incysted by way of electro-medication, or by their being administered internally to such person. In such cases the death of both would be the consequence.

Effects which can be produced by electro-magnetism through another can be made to imitate many established diseases which would receive the same diagnosis by the ordinarily instructed as well as by the learned physician. Their medicines would have no effect; the only remedy would be to disconnect them from the person partially incysted.

If a person should discover that he was connected with another by electro-magnetism and make such a statement, it would probably be discredited by the medical profession. They would give his condition the stereotyped diagnosis, " subjectiveness."

If he complained to his friends, or to his family physician, they would advise him to take rest and repose, and not permit his mind to dwell on the subject. If he complained to the officers of the law, he

would most probably be silenced by the suggestion that
he had better go before the Commissioners of Insanity.

THE BUDDHIST PURSUIT.

Not knowing how to disconnect himself from
others, nor how to de-magnetize himself, he can be kept
in that condition by persons unknown. This is called
the "Buddhist pursuit."

If he travels by land he may be transferred from
one jurisdiction to another, and be kept in the same
condition. The question is, can a person ever be de-
magnetized or disconnected from those with whom he
has once been connected?

SEMI-SUBJECTIVES.

In addition to the fact that the conscious force and
other forces may be partially united by a current of
electro-magnetism, by a proper use of the battery, the
chemicals, the electrodes, and the forceps, as hereinbe-
fore described, a system is practiced by which drugs,
such as opium and its alkaloids, etc., etc., are administered to one person, who is partially incysted in another, for the purpose of rendering the latter, or both,
in a condition similar to that of the subjectives, and inducing a portion of the conscious and sensory forces of
the latter to the former.

Let us represent two persons, one by C and the
other by D. Opium or its alkaloids or other drugs may
be administered by D, which, if given in the proper
quantity, will affect C to such an extent as to produce
sleep, whereupon, by an application of the proper elec-

trode of the battery to D (or to another person, who, not being affected by any drugs, may partially disconnect D and connect himself by the current of electro-magnetism in the place and stead of D), a portion of the conscious force and sensory forces of C, while in such a condition of sleep, under the drugs administered to D, can be induced to D or to the person who has taken D's place. In this condition C's thoughts are transferred. He may be questioned and cross-questioned as to his life, his acts, his opinions, his thoughts and his motives; his mode of expression may be by thought-transference or by a slight inner, inarticulated expression which may be interpreted and expressed by D or his substitute, or by another person connected.

Upon awakening C may have no recollection of what has occurred. What he does remember may be like the remembrance of a dream. Sometimes it may be faint, sometimes vivid. Unless he is aware of his condition of being connected with others by a current of electro-magnetism, he will relegate it to the realm of dreams. It must be remembered that it is not necessary to apply the electrode to C. A portion of C's conscious force is induced by the application of the electrode to D.

EXAMINING SEMI-SUBJECTIVES.

To those undertaking to listen, to question and to cross-examine persons in such a so-called condition of subjectiveness, care must necessarily be taken in separating the thoughts and expression of C's from those of D. If D's substitute attempts to interpret and express for C, the utmost care should be taken to avoid the interpretation of the mingled thoughts of C and those of D's substitute, as well as the natural tendency which D's substitute (or any person) has to anticipate what C is thinking about or wishes to say. Caution should be observed in D's substitute, as well as by the listeners and operators, to separate the thoughts and expressions, not only of C, and D's substitute, but the thoughts and expressions of others, whose brains may be to some extent connected by a current of electro-magnetism with C or D's substitute, or with both.

D having been disconnected from C may of course be connected by a current of electro-magnetism with some one else, and be questioned and cross-examined in a similar manner. (See a following paragraph on " Thought Transference " and " Mind Reading.")

Sometimes, a portion of the conscious and sensory forces of a person who is in a semi-subjective condition are partially incysted, first in one and then in another, his stray words or thoughts while in each, are often put together by the persons in whom he was partially incysted, or by their operators or listeners. Nearly all of the events of a person's life can be gathered by examination while in a semi-subjective condition; and pains are sometimes taken to do so, to make converts to spiritualism.

TRANSFER OF THOUGHT OF SEMI-SUBJECTIVES INCYSTED.

A person who is in a subjective or semi-subjective condition, whose conscious force and sensory forces are partially incysted in another, can be made to mentally express what a person who is connected with him wants him to express; or what the operator wishes him to mentally express, by the use of the electrode or any light electro-motive force, upon any person connected; and such person partially incysted can be made subject to all emotional excitement by impression.

DETECTING CRIMINALS.

This system of examination of persons in a condition similar to that of subjectiveness may be used to detect criminals (it has not to my knowledge ever been used by the civil authorities of any country for that purpose), to study psychological questions, to discover secrets, and it is said also to be used among theosophical sections for amusement.

SO-CALLED "TAKING POSSESSION."

The conscious force and the sensory forces of a strong person, may be carried or conducted with a current of electro-magnetism to a weaker one, and the weaker one be frightened into complete submission, being afraid to assert himself, particularly if he was not aware how the unison was affected—the stronger one might hot even know how the transfer was made; the current of electro-magnetism being sufficient, he might, as they say, "take possession" of the weaker one, and imagine that he had been completely transferred; in such instances there would be no element of fraud in either of the subjects; of course, the "taking possession" would be only temporary, and last while the current of electro-magnetism was induced.

DESCRIBING UNFAMILIAR LOCALITIES.

The optic nerves of A, being sufficiently connected with those of B, would enable A to describe localities, scenes, objects, paintings, etc., which he had never seen before, but which B was looking at, both being awake and not in any condition of subjectiveness. In other words, A would see through the eyes of B.

If the brains of two persons were connected—let us represent them by C and D: if C was asleep or in a subjective state, or in so-called subjective state, the operator could direct him to visit a certain locality, this would be communicated to D, by C repeating it to D by inarticulated expression, as in reading, or by the fact of C hearing it, by reason of being connected, so that D could visit the locality named, and describe it to C by the same method of expression, or by thought transference; the object which D looks at is expressed in his mind and transferred to the mind of C, if you look at a landscape, a tree, a chair or a book, by the very act of observation the object is described.

C having been in a subjective state, or, we will say, asleep, if enough of his conscious and sensory forces are induced to D, C will conclude that he has visited the locality described, unless he knows better, by being aware of having been connected with D by a current of electro-magnetism.

INCYSTING AT GREAT DISTANCES.

A person's conscious force while in a semi-subjective state, may be partially incysted in another, who may be at such a distance that the person incysted could with truth say that he had seen the sun rise in the night time.

A natural current of magnetism, passing from one person to another, by the earth's current, may connect the conscious and sensory forces of persons at great distances apart, where it was not purposely effected.

A person who is in a semi subjective condition, whose conscious and sensory forces are partially incysted in another, may be so connected with others who are not subjective, that they can question him as to where he is, what he sees and what he hears.

TELEPSYCHOLOGICAL ACTION.

THOUGHT TRANSFERENCE—MIND READING.

The origin, development, transfer, interpretation and expression of thought, by persons whose brains are connected by a current of electro-magnetism, are matters of the utmost importance and deserve careful study. The thoughts of another, or others, having been transferred, are often taken to be the original thoughts of the person to whom they are transferred. This would always be the case with the person receiving them, if he did not know that his brain was connected with another or with others, and thus give himself credit for great originality.

A thought of one person often results in its expression by others. After a thought, as we may say, is in embryo, its subsequent development into articulated speech, assumes that form of expression, such as we use in selecting words, or such as we use in its expression from the mind to the pen.

Among persons whose brains are so connected, difficulty is often experienced in detecting the originator of a thought in embryo, its transfer in that condition to another who attempts to develop it, is often misinterpreted by reason of the natural tendency which every person has to anticipate what another is going to say.

The thought of one may be almost contemporaneously joined with that of another, and both seem to originate in the same person, which, being transferred to others, are, in consequence of their seeming to issue from the same brain, misinterpreted, misconstrued and given a false expression. The same difficulty arises when a part of a sentence is mentally expressed by the person originating it, and other mental words are transferred to him from another at or about the same time; the " part of the sentence " and the " other mental words " passing to others and coming apparently from one person will most generally fail to be separated by the others, and consequently an unmeaning phrase or a false expression will be given, there will be a misjoinder of words.

The person interpreting should have a fair and an honest mind, a thorough knowledge of character, and should be careful not to mingle his own thoughts with those transferred, as well as not to anticipate.

A person whose brain is connected with the brains of others by a current of electro-magnetism may be thinking of a person thus connected with him, and by concentrating his mind upon him, convey that impression to such person, who will mentally respond to it, even if he is pursuing his usual avocations; this is more readily accomplished by associating the person thought about in the first instance with some locality where they have both been, which made a strong impression upon their minds by the fact of their having often been there together, or by the fact of some interesting or unusual incident having happened there. The person on whom the mind is concentrated and the will power of another is exercised, may, by fixing his attention. if he so desires, receive the thoughts and transfer his own. and by a proper application of the pole of the battery to his head, (or by stronger current being generated by some one else who is also connected) induce more of his sensory forces to such person, so that the transfer of thought becomes more and more expressive.

Thought of others may be passed through a certain part of the brain of one or more persons connected without their knowledge, such persons acting as conductors only.

MENTAL WORDS AND WORDS IN THE OUTER CURRENT OF ELECTRO-MAGNETISM.

Words are often framed in the mind even by letters; they can sometimes be seen, as it were, in the atmosphere, as it were, imprinted in the outer current of electro-magnetism which connects one person with another.

USING THE SERUM AND MAGNETISM OF OTHERS — VICTIM SYSTEM.

The serum, or some of the substances of which the serum is composed, or the animal magnetism, has the potentiality of thought and expression when it is induced to another; it can be used by him to a great extent in the same manner, as it can be used by the person from whom it is induced.

A person may change his method of thought and expression by inducing to him the serum or magnetism of another.

Qualities of mind which may be lacking in one person may be induced from another, such as reason, judgment, mental activity and energy, the emotions and the temperament. How long the qualities of mind so induced will remain I am not prepared to state.

Some portion of the serum or magnetism is needed for exact expression; when a portion of it is induced to another, it will partly return to the owner by the exercise of his will power; by the very effort he makes to express himself, by articulation or talking, a portion of the serum or magnetism returns. The action of the muscles used for inarticulated and articulated expression by the will to express a thought, causes the return of a part of the serum or magnetism. What he intends to say, however, may be interrupted " and cause him to forget what he was going to say " by a person whose brain is connected using the electrode on himself. The greater the amount of electro motive force used to interrupt, the greater must be the effort made for the expression of thought.

A thought in an embryonic condition, is first, what I may term a mental sensation, it afterwards becomes a physical one by a sensation of the brain, and becomes more and more pronounced by its action on the nerves, until it takes the form of articulated or written expression. When the serum or magnetism is partly induced to another, it is difficult for the person from whom it is induced to develop thought into mental words, even if not interrupted; it requires an inarticutated. or written, or articulated expression of the thought, which. when expressed. returns to the mind, gives mental satisfaction and fixes it in the understanding.

A thought in an embryonic condition, or when partially expressed, may be taken involuntarily from another by the application of the electrode to the head of the person with whom he is connected.

In addition to the power which has been exercised of using the brains of others. as it may be termed, for their qualities and characteristics, the serum and magnetism of the body as well as of the brain, are often used for their vitality for the purpose of recuperating wasted energy.

This probably explains the victim and vampire system of India and other countries.

It has been asserted by some persons accustomed to have the serum, or substances which compose it, or magnetism of others induced to them, that they can identify the person from whom it came. This, however, is very doubtful, as the serum or magnetism of one may not only have the same peculiarities or qualities as that of others, to some of the senses, but it may be so mixed with that of others that it cannot be identified.

RAPID COMMUNICATION WITHOUT TELEGRAPH.

" Sporting men are not very useful people, but we " are inclined to think they could just now perform a " small service for the world, by clearing up a problem " which every now and then has perplexed soldiers.

" statesmen and historians. How do Asiatics without
" telegraphs, or semaphores, or heliographs—though as
" they have mirrors and ingenuity they ought to have
" invented these latter—contrive to transmit the heads
" of intelligence so rapidly as they do?

"The Crusaders, so far as our reading extends.
" either never observed the fact or they were not sur-
" prised at it; but ever since Europe entered Asia
" generals have noticed that accurate rumors of
" startling events have been known to the natives
" around them before they themselves received the in-
" formation.

" The story as told by the dark men was usually
" overlaid with details manifestly false, but the central
" kernel often or even usually turned out true. They
" have observed that the news arrives in some form
" that appears to Asiatics trustworthy, for its recipients,
" whether enemies or traitors, or only large dealers on
" the 'Change, have acted unhesitatingly, often staking
" on its accuracy either their fortunes or their heads.

" The writer was himself informed in Calcutta of
" the defeat of Chillianwallah two days before it was
" known to the government, and knew of the bare fact
" of the redeeming victory of Gujavat twelve hours
" before the departments received the intelligence. The
" latter was a most remarkable case, as the government
" had made special arrangements to secure early in-
" formation along the whole line, and were vexed at
" their defeat, and never accurately ascertained how
" their messengers had been outstripped."

London Spectator, February 21st, 1885.

Thought transference most probably explains the sys-

tem of rapid and secret communication in India by
some writers termed the " secret mail service."

DISTANCE.

Various estimates have been made of the rate of
speed at which the nervous force travels, so we will be
able to measure the rate of passage of thought from
brain to brain separated by the distance of the earth's
diameter.

THE WILL POWER.

The exercise of the will power is illustrated by the
machine called a " reel," with magnetic needles, invented by Count P. (who has sought a retreat near
London in order to avoid those who, under pretext of
scientific inquiry, merely seek to derive amusement
from the most serious experiments in science), who
pursued many of his experiments with Ruhmkorff. An
imperfect description of the reel in question has been
published in the current journals. It seems that " without speech, without touch, by the mere mental influence
alone will the machine move in obedience to the unex-
pressed demand."

TELEPATHY.

The emotional glands of one person being excited
will produce a degree of excitement in the corresponding glands of another, where they are connected with a
current of electro-magnetism.

TRANSFER OF IMPRESSIONS.

The operator can apply the electrode to any part of
his or the subject's head, and if connected with another

or others, can by thought, transfer to them any subject or impression, including all emotions which he is capable of expressing, feeling or simulating.

Thus persons may be impressed with feelings of irritability, anger, hatred and revenge: with contempt, disgust and nausea: with melancholia, hypocondriasis: gloom, despair, with homicidal and suicidal inclinations; on the other hand with impulses of forgiveness, friendship, sympathy and affection or respect, with feelings of repugance or amativeness; the thoughts can also be directed to devotional worship, to the contemplation of a future life, and to all ontological and philosophical speculation.

All persons connected in one or all of the magnetical unions, can be made to receive one and the same impression. Even those who know they are connected with others, and understand how they can be impressed with emotions, often fail to consider whether they are being impressed or not, by another.

The many, who are not familiar with the method of transfer of impressions, often mistake their impressions for their opinions. In this way a whole community may lose their independence of thonght, and their actions may be guided by a galvanic battery. They may be prejudiced against one individual, against a race or a religion.

MUSCLE READING.

Words, of course, may be conducted from the brain through the nerves to the muscles, and may be expressed by sound. Thus, if a word is in the mind, and attention is given to its instantaneous transmission from the brain to the foot, and from the foot to the floor, the sound produced by the stamping of the foot upon the floor will express the word. This method of expression is limited to short words; a long word may be expressed in the same manner by dividing it into syllables. By striking a resonating metallic object with metal or a cane, adopting the same method of instantaneous transfer, the sound can be made to carry the syllables and words.

In shaking hands the same principle can be applied.

Some persons who are in some condition of electromagnetism are familiar with this method of communication, and can readily translate it. Others can even translate gestures with the hand, or with a cane which is made to carry words from the mind.

Every group of muscles which can be easily moved or controlled by the will. can be made to express inarticulated words.

INARTICULATED EXPRESSION BY THE FACIAL MUSCLES.

A current can be made to pass through the facial muscles of a person without his knowledge, inarticulately expressing whatever the operator wishes. This is done by the operator using the electrode on another whose facial muscles are connected with such person. Persons who are in certain conditions of electro-magnetism read what is expressed in a manner similar to that used by deaf people reading the muscular action of the lips of a person talking.

MISTAKES IN MUSCLE AND MIND READING.

Facial muscles, and movements of the lips, may seem to convey words to a person reading inarticulated expression, while in fact it is an effect produced by the operator upon the person reading, the thought or mental words come from the operator and pass through his subject who is connected with the person reading, the latter attributes them to the person he is looking at, studying, or.|as it may be termed, mind or muscle reading.

The muscles of a person, or a particular muscle, may be so electro-magnetized with a word or a sentence, so that it is expressed by every movement of the muscles and can be translated by others who are in a condition of electro-magnetism; this condition of the muscles being charged to express a word or sentence may last for a considerable time, repeating with every movement the particular word or phrase.

But little credit should be given to thought transference, or to any of the systems by which inarticulated expression is conducted.

VOICE TELEGRAPHING.

The natural voices of persons in a condition of electro-magnetism, or the voices which they transmit, can be heard at considerable distances, by the auditory nerves of the ear having been sensitized by the application of a current from the galvanic or other battery; or perhaps when so sensitized by certain *drugs or chemicals*. Voices may be heard three or four miles, perhaps several hundred. It depends on the condition of electro-magnetism of the person speaking and the person hearing, respectively, how they are connected, and the substances, chemicals or ethers used.

No reliance can be placed upon any identification of a voice, as it is a common practice to imitate voices; it has been stated that one family who have been electro-magnetized for many years, have imitated over two hundred voices of residents of San Francisco, both male and female.

CHANGE OF VOICE.

A person's voice may be changed by uniting the organs of articulation connected with another, so may his laughter.

INARTICULATED WORDS RENDERED ARTICULATE.

An operator can pass words and sentences through another who is engaged in a conversation, what the operator wishes to say through the person speaking is conducted from another, through the thorax or mouth, in the form of inarticulated expression, which becomes articulated by the vibrations of the voice of the person speaking. Of course the inarticulated words must reach the mouth of the person speaking at the instant of his articulated expression.

A person in a condition of electro-magnetism can listen to both.

Sometimes it is difficult to distinguish the words which are spoken from those which are transferred or conducted. In this way two subjects of conversation can be carried on at the same time, apparently from the same person; one, however. comes from the operator. A person with practice can listen to both: or one person may listen to one, and another to the other.

Persons may often be mistaken, as to what was really said under such circumstances. The operator may change a negative into an affirmative to the person listening, while the person speaking may be ignorant of what was transmitted or conducted through him.

Humming or whistling the words is another secret method some persons have of communicating with each other ; under electro-magnetism, it is probably easier to hear and understand what is hummed or whistled.

PHONOMANIA.

This term I first used as applicable to words and sentences, or to speech and whispers said to be heard by some persons, and called hallucinations or illusions of hearing, "due to derangement of the perceptional ganglia"—if it is true that there are any such hallucinatons.

There may be such a condition of the brain, that unusual sounds are said to be heard, such as buzzing. hissing and ringing sounds; but I do not believe that when words are distinctly heard, never mind how often they are repeated, that they are illusions, but, on the contrary, are the utterances of living human beings, who may be at considerable distances, in the same neighborhood or in the same city, especially when the person hearing them is certain that what he hears does not emanate from his own brain.

Phonomania may also be illustrated by what Mark Twain says in his " Sleeping-Car Experience ":

" One night, having raised my window curtain to look over a moonlighted landscape, as I pulled it down, the lines of a popular comic song flashed across me. Fatal error! The brain instantly took it up and during the rest of the night I was haunted with the awful refrain. * * * "

AUGMENTATION OF SOUND.

Motion and sound augment the voices of persons who are in a condition of electro-magnetism to others in a like condition.

VISUALIZING.

It is well known that many persons have the faculty of mind-picturing, and can by an effort outline the faces of persons they have once seen; others lack this power. The power of forming pictures of objects in the mind's eye is termed by Francis Galton, "visualizing." It is said by him to be a natural gift, and like all natural gifts, has a tendency to be inherited, and exists in a higher degree in the female sex than in the male.

Some artists can paint from memory, others find it difficult, and some find it impossible. It is related of a painter, who had painted three hundred large and small portraits during the course of one year, that it was his custom to look attentively at his sitter for about half an hour, sketching from time to time; he would then remove the canvas and sketch another person for about the same period of time; when he wished to continue the first portrait, he recalled the man to mind as sitting in the chair which he had previously occupied, where he could perceive him as distinctly as if he were really there, in form and color more decided and brilliant, and went on painting, looking from time to time at the imaginary figure.

Horace Vernet was celebrated for his power to paint from memory.

It is related of Talma, the great actor, " that he could, by the power of his imagination, cause the audience to appear like skeletons, and that when the hallucination was complete his histrionic genius was at its height."

Goethe states that he had the power of giving form to the images passing before his mind.

Galton, speaking of persons who have the "visualizing faculty," says: " Others have a complete mastery " over their mental images; they can call up the figure " of a friend, and make it sit in a chair or stand up at " will; they can make it turn around and attudinize in " any way, as by mounting it on a bicycle or compell- " ing it to perform gymnastic feats on a trapeze." An eminent mineralogist assured Galton that he could imagine simultaneously all the sides of a crystal with which he was familiar. (*Fortnightly Review*, September, 1880.)

Galton suggests the cultivation of " the capacity of " calling up at will a clear, steady and complete mental " image of any object that we have recently examined " and studied." He says: " We should be able to visu- " alize that object freely from all sides; we should be

" able to project any of its images on paper and draw
" its outline there. * * * I believe that
" a serious study of the best method of developing the
" faculty of visualizing is one of the many pressing de-
" siderata in the new science of education." He, however, says that " *when the faculty is strong it is apt to run
riot.*" Like every other operation of the mind, it should
be controlled by the will.

Many persons in some conditions of electro-magnetism acquire the faculty of mind-picturing readily and
without instruction; by the least effort of thought they
can picture to themselves not only the faces which they
have once seen, but faces unknown, such as artists paint
for their studies; they can also picture objects, un-
sightly figures, monsters of all shape and mephistophe-
lian forms; such mental pictures will often be conveyed
to the minds of persons whose brains are connected by
a current of electro-magnetism. So, one person may
picture a word or an object, and another person rub it
out; one may give to an imaginary person a certain position, and the other may alter it, and thus, as it were,
have a visualizing combat.

The effect which a current of electro-magnetism has
upon the natural phosphorus of the brain may in a
measure account for the facility of visualizing.

Persons in a condition of electro-magnetism should
be careful not to practice visualizing too much; there
might be some danger of their losing all power of dis-
tinction between the imaginary and the real figure.

Visualizing by women who are in a condition of
pregnancy should be avoided. The effects of visualiz-

ing by the mother during pregnancy, or visualizing by others who are connected with her by electro-magnetism while in such condition, may account for many of the terrifying hallucinations of sight, as well as in some instances for the birth of monstrosities.

CHANGE OF FEATURES AND EXPRESSION.

A person's features and expression may be temporarily changed by passing a current of electro-magnetism from another to him, so that his identity could hardly be sworn to by his friends, without a close inspection.

COMPOSITE FACES.

Faces can be temporarily altered so that they bear a resemblance to two or more persons, a different expression and tinge can be given to the eyes, and the complexion may be discolored.

These changes of features and expression can be made by a skillful operator using the electrode on the facial muscles of another person connected.

Whether persons who are not in any condition of electro-magnetism can observe these effects I am not prepared to state.

TRANSMISSION BY SIGHT.

If A, or any person connected with B, or any person connected with A, and thereby with B, by a current of electro-magnetism passing through their cerebrums and cerebellums sees an object, such as the painted name of a street on a sign, a landscape or other scene, the object or the landscape seen may be reflected to B,

passing through the optic nerves of the persons seeing
it, making its temporary impression on the brain,
may sometimes be seen in the cerebrum of B. with his
eyes *closed*, also it may be sometimes seen apparently
before the closed eyes of B. This is done by a skillful
use of the electrode, sometimes with the aid of the
rheotome. In such cases the object is reflected, pro-
viding that the optic nerves of A, or those of the person seeing the object, happen to be in the proper line
of electro-magnetism with B; perhaps the person could
be so placed, and in that way objects might be purposely reflected from one person to another.

A person, seeing an object, its spectrum, that lin-
gering impression made by it upon the eyes, may sometimes be reflected to the open eyes of another.

In order that one person should read through the
eyes of another, it is necessary to induce more of the
conscious and sensory forces from one to the other, and
the optic nerves of both must be in unison. ·

No wonder " there is a blind man in Washington
who can see!"

EFFECTS OF PRE-NATAL ELECTRO-MAGNETISM.

What may be the effect upon a child conceived and
born while one or both of the parents are in a condition
of electro-magnetism is a subject deserving careful investigation. Perhaps the child, especially the female,
might inherit peculiar psycho-physical conditions with
relation to its parents, especially to its mother.

Another subject for investigation is whether a child
born in such a condition shows a greater tendency to

subjectiveness, and whether it is more easily electro-magnetized, mesmerized or hypnotized than others, and how far all children born of such parents are affected.

HALLUCINATIONS OF FEVERS.

The condition of the brains of persons who are suffering from fevers and are " out of their minds," as it is termed, should be carefully investigated with reference to the condition of their auditory nerves: whether, the brain becomes in such state that it is capable of receiving mental sounds, and whether they can hear the voices of magnetics.

It is well known that persons who are in the possession of the secret of the Buddhists and their subjects or victims, and others who are in a condition of electro-magnetism, meet during the unusual hours of the morning to hold their nocturnal rites, their frantic revels, bacchanals and orgies.

The Druids sometimes meet at midnight and enact rows, combats, and pretend to murder, for the purpose of fooling their neighbors and enjoying a good laugh the next day at their expense, while some of the Spiritualists, both male and female, play the part of the insane, until one would think all Bedlam had been turned loose.

Others hold representations of scenes which never happened, taking the parts and assuming the names, and imitating the voices of others for the purpose of spreading false rumors and scandal.

As persons in a condition of electro—magnetism can hear one another at considerable distances, perhaps persons whose brains are in fever can do the same.

The fright. wanderings and hallucinations of such patients may be due to such cause.

Whether the brain becomes in that condition by the over use of alchohol, bromides. opium, morphine, chloral, etc.. should also be investigated.

If the hallucinations of a fevered patient should prove by investigation to be due to hearing these natural voices, an explanation to the patient would be more effective than the medicines.

EFFECT OF MEDICINES ON PERSONS ELECTRO-MAGNETIZED AND CONNECTED.

The rapid and immediate effects which many drugs and medicines have, when administered to persons who are in a condition of electro-magnetism, and their transmission to others who are connected, has probably been investigated by physicians who understand the subject. the result of their observation should be made known to their *confreres* who have been kept in ignorance of electro-magnetism. For the purpose of experimenting. the introduction of medicines into the system of persons in such condition by the galvanic current deserves some attention.

HYPNOTISM AND MAGNETISM.

A recent work by Frederick Bjornstrom, M. D.. of Stockholm, translated by Baron Nils Posse, M. D.. Director of the Boston School of Gymnastics. on " Hypnotism; Its History and Present Development," gives a historical retrospect of the subject of magnetism. He frequently uses the term hypnotism. indicating a condition akin to magnetism. Like many others, who use the terms magnetism, mesmerism, hypnotism and spiritualism as synonyms. Bjornstrom refers to the introduction of mesmerism in France; to the secret magnetic order of Mesmer's day*; to the famous " Harmonic Society " (Societe de l'Harmonie), and to La Societe de la Guyenne, and to the vast magnetic league, which in the present day works so much mischief in Paris.

He fails to mention the Harmonic Brotherhood of Saxon, an old magnetic league; perhaps it has ceased to exist.

He alludes to the dangers of hypnotism, and by way of advice, he prefaces his work with the ancient · classical dictum, " Observandum sed non imitandum."

He describes the various phenomena of hypnotism, such as Transmission of Sensation, Transmission of Images. without words or signs, without the aid of the external senses, wholly mentally, and to the " stigmata produced by means of hypnotism, without deceit and without the miracles of higher powers." He only describes these phenomena, he gives his authorities, and the time and place of their occurrence; he does not account for them except by his reference to magnetism. This ought to explain all; it is the magnetizing of the human system by the galvanic battery, and does not mean hypnotism, nor mesmerism nor spiritualism.

In the present day many attempts have been made to expose the secret of magnetism, but the very use of

---

Mesmer, it is said, refused 340,000 livres to reveal his secret.—*Chambers' Encyclopedia.*

the terms employed, such as hypnotism, etc., have confused and misled the public.

The better authorities say that few persons can be hypnotized or mesmerized by the will of another, without resorting to the usual method of tiring or affecting the optic nerve by long concentration on a brilliant object.

In describing hypnotic phenomena proper, it is absolutely necessary to discard all those cases described under the head of hypnosis, where the subjects have been electro-magnetized and connected with others.

A reference to the works of Count Rabiano, a French abbé, mentions that all the effects of somnambulism can be reproduced by the galvanic current.

Most of the subjects, or sensitives as they are called, have probably been electro-magnetized, without their condition being made known to them; their will power has been enfeebled, they have not been allowed to use it, having been made to yield complete obedience to the person who has had them in custody, or to the voice which they have heard at a distance, which they have been accustomed to obey, or to a command given by transfer of thought or to an impression.

Where they have sought to assert themselves or rebel, they have probably been frightened into complete submission.

BIOLOGY.

The effect of electricity and magnetism upon the beginning and development of life will afford a wider field of study and investigation to biologists when it is better understood.

DREAMS.

The unison of the conscious force, may by investigation throw some light on the subject of dreams. Every person who is asleep is in a semi-subjective condition. The vividness of some dreams which have been described by persons who were not connected with others by a current of electro-magnetism, would lead us to believe that if a portion of their conscious force had not been induced to others, by a natural current of magnetism, they were in a condition to hear what others said and imagined they were taking a part, in a conversation.

INSANITY.

Let this question of electro-magnetism be investigated by our physicians. It is the most important question to which their attention has ever been called. It will account for many " cases " which they have failed to diagnose and treat; it will throw light on those cases which have puzzled the most learned; it may explain the so-called tendency to insanity, and account for many neuresthenic diseases. In fact its investigation and study will be a new era in the history of medicine.

IGNORANCE OF ELECTRO-MAGNETISM.

The application of electro-magnetism to the human body has, it is said, been known for centuries. Among all nations, it is the secret of Buddhists, and is known in many schools of philosophy and among a number of Sects, Tribes and Orders, and among the *illuminati*. It is one of the secrets of Mecca and Constantinople, and it is said to have been sealed in the Vatican. Most of the remarkable phenomena produced by the magicians of India may undoubtedly be accounted for by their knowledge of magnetism. It is undoubtedly through it that the Kabal prophets and seeresses were developed.

It is probable that music heard and transferred to others at great distances was the " harmony of spheres " listened to by the ancients.

The application of electro-magnetism in the manner described has been kept a secret from the mass of mankind. which unfortunately has been the cause of spreading superstition and increasing the belief in spiritualism, and has been. perhaps. the cause of other evils which have never been attributed to it by reason of ignorance and lack of investigation.

The mind of man should be relieved from its inherited and nurstled superstitious tendencies. and should be taught to account for all phenomena by natural laws.

The literature on the subject is said to be rare. Most of the editions of works describing or alluding to its use have found their way into inaccessible libraries. In one of the rarer editions of Paracelsus there is said to be a short description of how magnetism is applied to the human body. It is probable that the most important writings on the subject have never been published. and can only be found in the archives of the initiated.

MAGNETISM —THE BUDDHISTS' SECRET.

So careful have the possessors of the Buddhists' secret been, that in general literature nothing can be found upon the subject of magnetizing the body by electricity. No reference is made to its dangers; but little reference is made to the cause of the magnetical wars of Europe and Asia.

We find references to mysticism, to black magic, to wonder workers, but no explanation of how the wonders are wrought.

DANGERS OF MAGNETISM AND SECRET SOCIETIES.

In the history of Druidism we are told that Pythagoras taught them magic and the doctrine of the transmigration of souls, and was the founder of their philosophy; he in his turn was instructed in the mysteries of Thebes and Egypt.

He formed a secret society and no one was admitted except after severe initiation. The novice was condemned to years of silence, to various humiliations, to self-denials and to trials of endurance.

" He (Pythagoras) refused, at least for a time, os-
" tensible power and office, and was contented with.
" instituting an organized and formidable society, not
" dissimilar to that mighty order founded by Loyola in
" times comparatively recent.

" The disciples admitted into this society underwent
" examination and probation; it was through degrees
" that they passed into higher honors and were admitted
" into deeper secrets. Religion made the basis of the
" fraternity, but religion connected with human ends of
" advancement of power.

" He selected the three hundred who at Croton
" formed his order from the noblest families, and they
" were professedly reared to know themselves that they

" might be fitted to command the world. It was not
" long before this society, of which Pythagoras was the
" head, appears to have supplanted the ancient Senate
" and obtained the legislative administration ! "

＊　　　＊　　　＊　　　＊　　　＊　　　＊

" An order based upon so profound a knowledge of
" all that can fascinate or cheat mankind could not fail
" to secure a temporary power.   His influence was un-
" bounded in Croton; it extended to other Italian cities;
" it amended or overturned political constitutions.

＊　　　＊　　　＊　　　＊　　　＊

" It was when this power so mystic and so revolu-
" tionary had, by the means of branch societies. es-
" tablished itself thoroughout a considerable portion of
" Italy that a general feeling of alarm and suspicion
" broke out against the sage and his sectarians.   The
" Anti-Pythagorean risings, according to Porphyry, were
" sufficiently numerous and active to be remembered
" long generations afterwards.   Many of the sage's
" friends are said to have perished. and it is doubtful
" whether Pythagoras himself fell a victim to the rage
" of his enemies or died a fugitive amongst his disciples
" at Metapontum.   Nor was it until nearly the whole
" of Lower Italy was torn by convulsions and Greece
" herself drawn into the contest as pacificator and arbi-
" ter that the ferment was allayed."
History of Philosophy.— Lewes.

" He was a worker of miracles.   He was heard to
" lecture at different places, such as Metapontum
" and Taurominium on the same day and at the same
" hour.

This is one chapter in the history of magnetism, but enough to illustrate its danger.

MAGNETISM AND NIRVANA.

Now let us consider this subject of electro-magnetism in connection with the Nirvana of the Buddhists.

Druidism, which may be said to be an offshoot of Buddhism, teaches that men's souls do not perish, but transmigrate after death from one individual to another (they evidently must have believed that it is accomplished by magnetism), and they hold that people are thereby most strongly urged to bravery as the fear of death is thus destroyed. The Buddhists evidently do not contemplate this system of absorption with pleasure; their hope and prayer is the final deliverance of the soul from transmigration. They seek rest or even complete annihilation in preference.

The use of another's serum or some of the substances which compose it, and the animal magnetism and their vitalizing effects, before referred to, under the title of " Transfer of Magnetism," will suggest a method of determining this question of absorption and transmigration mentioned in Druidal philosophy, by scientific experimentation.

ROME CONDEMNS MAGNETISM.

The Church of Rome has frequently condemned magnetism. Late dispatches (September 5th, 1890) indicate that the Holy Inquisition condemns Hypnotism upon the ground that *it disturbs human liberty* and is dangerous in its effect upon the mental and physical condition of hypnotized subjects.

EXPOSURE—THE SANSUVAH.

To have the Buddhists' secret of magnetism exposed. has been the hope and prayer of millions of souls for century after century. All attempts have failed; the exposers have generally met with persecution and death. Persons who have re-discovered the secret have usually been persuaded or coerced into silence.

Since the year A. D. 1777, the Sansuvah has been convoked almost every decade. At nearly every assemblage it was agreed that Buddhism, or in other words. spiritualism or magnetism should be exposed, not only to scientists, but to the whole civilized world.

The difficulties in the way of exposure, in addition to forfeiture of large sums of money, secured by bonds and estates of leading financiers, have been numerous.

It involved a confession of fraud and protection of crime; secrets of state and church were known to the different representatives in the Sansuvah respectively, which, though protected by promise as well as by fear. were threatened to be revealed by those opposing it.

Its exposure involved the mysteries and rites on which religions were founded; from the days when the oracles of Delphi and Apollo foretold the success or destruction of armies; the rise and fall of nations. magnetism has been protected by the arms of kingdoms and the superstitions of mankind.

Altho' nations have disappeared, and their history is only to be be found in the sarcophagi, and recorded on papryus, or inscribed on as yet undeciphered syenites and basalts. Magnetism and its secrets, have been handed down from generation to generation as the

Buddhists have sung the history and hymns of their religion to their children.

In the Sansuvah, the leading religions of the world have been represented, though not equally; Buddhism, The Church of Rome, The Greek Church, Brahmahism and Judaism. In A. D. 1846 the Roman Catholic Hierarchy withdrew, and since that day they have not participated in the proceedings of the Sansuvah. In St. Peter's, where Michael Angelo left the inspiration of his genius, and converted a saracenic hall into a Christian church; where music echoes a prayer, the Church of Rome has appealed for protection against magnetism.

Who is responsible for subjecting a large majority of the residents of San Francisco to electro-magnetism? It is needless to further allude to their condition; many of them do not recognize it.

It has been asserted that when a part of the body or some of the muscles have been connected with another, the rest may be.

A person may be connected with another through a group of muscles, or through the brain, for years, without knowing it; many do not discover it until they begin to hear voices of other persons in a like condition, at long distances. They should be informed.

I comprehend that once upon a time the use of magnetism for punishment was limited to those cases, where the laws of the country were inadequate.

It has, however, been used for unjust punishment and persecution. *The proper time has come for its exposure.*

S. HEYDENFELDT, Jr.,
September, 1890, San Francisco, Cal., U. S. A.

.

APPENDIX.

DISTRIBUTING MEDIUMS.

I understand that the following is the method of
keeping persons in a condition of electro-magnetism:

In order that persons should continue to be connected with others, after having been electro–magnetized and connected, and remain in what is called a
magnetical union, one or more persons are made distributing mediums or subjects, with whom others have
been connected by the galvanic current or by the use of
faradic and medical coil batteries; to them are administered food, chemicals, drugs, etc., which generate in
the serum, substances (ethers by way of electro-medication) which aid in continuing the current, which is made
to pass from one to the others; *cerium* is said to be an
important factor. Gum arabic and other hydro-carbons
are · taken internally, and are also held in the mouth
of the distributing medium and allowed to dissolve
slowly.

PERSONS IN IMMEDIATE CONNECTION.

Of course a number of persons are immediately connected with the distributing medium, on whom the current acts directly, passing through and from such
medium to such persons, and through them to others.
By increasing the quantity of electro–motive force, the
brains, muscles, tissues, fibres and nerves of all can be

connected to a greater extent with those of the distributing medium.

The persons immediately and directly connected with the distributing medium, assist as conductors, the current from the battery passes through them to others, carrying substances, serum, or some of the substances which compose the serum, magnetism, ethers, etc., from the distributing medium, as well as the substances. magnetism and electricity which they generate in themselves. Some of them carry with them and occasionally use a small galvanic battery, or medical coil battery. The earth's current passes through them all.

DIRECT CONNECTION OF THE DISTRIBUTING MEDIUM.

The current of electro-magnetism, therefore, passing from the distributing medium, carrying and conducting with it the serum, or some of the substances which compose the serum, ether, etc., as well as the magnetism from each person, connects the brain of the distributing medium *directly* with each and every person in the magnetic union. By the skillful use of the electrode one person can be more closely connected and the others almost disconnected.

WHEN ONCE CONNECTED.

When persons have been once connected with one of the distributing mediums, they continue to remain subject to electro-magnetism, as long as a light current, imperceptible to most of those connected, is applied to the distributing medium. How long this connection will remain after the disuse of the battery, is for those familiar with the subject to expose.

CLOSER CONNECTION—GREATER DEGREE OF ELECTRO-MAGNETISM.

When a person is connected with the distributing medium by such a light current that he is not aware of it, and it is desired to electro-magnetize him more or to connect him closer, or induce more of his sensory forces to the medium, it is necessary to increase the quantity of electro-motive force. To do so the distributing medium, or his operator, by concentrating his thought upon the person at the time of using the electrode, or by looking at a photograph of such person and fixing his mind upon him, at the time the current from the battery is applied, causes the current to pass from the head of the distributing medium to such person thought about, and establishes a nearer and more immediate connection.

Others will have to describe the larger batteries which generate the electro-motive force and the general electrical mechanism and also the chemicals kept in rubber sacks, distributed in different parts of San Francisco by which thousands of persons are electro-magnetized and kept in that condition.

ELECTRO MAGNETISING AND RE-ELECTRO MAGNETISING.

Whether a person can be electro-magnetized, or be re-electro-magnetized by a distributing medium. through the solar rays and a looking glass, is a question for magnetizers to explain.

Electro-magnetizing may have been facilitated by using a current of electricity in connection with the street cable cars of San Francisco.

50

A NUMBER OF MAGNETICAL UNIONS.

There may be a number of these distributing mediums; each may be connected with the other; in consequence, the persons who have been connected with each of these distributing mediums are liable at any time to be electro-magnetized by any of the distributing mediums or by their operators.

In order to keep people subject to electro-magnetism, a current of electricity is applied every now and then upon these distributing mediums, most of whom are generally kept unidentified and secreted.

The application of the galvanic current upon the distributing medium will produce somewhat of a corresponding effect upon the persons connected with him, though to a less extent; the faradic likewise.

SIGHT OF THE DISTRIBUTING MEDIUM.

A distributing medium can be so connected with many others that he can see what they see; this must be left for his or an operator's scientific description. There is said to be a system by which persons and objects which are seen by persons connected with the distributing medium, can be reflected in a mirror before him.

The effects and phenomena which can be produced by electro-magnetism and chemicals, are so numerous that it would take volumes for their description. Many schools have their own secrets, which they try to guard most carefully from others. These schools have their skilled operators and chemists; each tries to out do the other (even by examining their electricians and magicians under hypnosis). There is a constant rivalry; they are jealous of the one who can produce the newest and most unaccountable effects; therefore new discoveries are being made periodically, which all seek to imitate and improve upon.

The use of luminous ethers and electro-magnetism by the Theosophical Sections, will probably be published to the scientific world.

The magnetico-chemical figures seemingly imbued with life, and the Eleusinian and Egyptian mysteries will probably be explained by another Order.

In the foregoing pages, many heretofore unaccountable phenomena have been explained—enough, it is to be hoped, to direct the public mind to a solution of all others by science and natural laws, and to discard forever superstition and spiritualism.

S. HEYDENFELDT, Jr.

San Francisco, Cal., September, 1890.